D1288922

Extreme Animals

Nature's Deadliest Animals

Frankie Stout

PowerKiDS press.
New York

For Nicholas Anthony Lazarus, a wonderful nephew

Published in 2008 by The Rosen Publishing Group, Inc.
29 East 21st Street, New York, NY 10010

First Edition

Editor: Jennifer Way
Book Design: Greg Tucker
Photo Researcher: Nicole Pristash

Photo Credits: Cover, pp. 5, 9, 14 (inset), 17 Shutterstock.com; p. 7, 19 © Ian Waldie/Getty Images; p. 11 © SuperStock, Inc.; p. 13 © Michael & Patricia Fogden/Getty Images; pp. 14–15 © National Geographic/Getty Images; p. 21 by Artville.

Library of Congress Cataloging-in-Publication Data

Stout, Frankie.
Nature's deadliest animals / Frankie Stout. — 1st ed.
p. cm. — (Extreme animals)
Includes index.
ISBN 978-1-4042-4159-6 (library binding)
1. Dangerous animals—Juvenile literature. I. Title.
QL100.S766 2008
591.6'5—dc22

2007026723

Manufactured in the United States of America

Contents

Deadly Animals

Animals big and small can hurt and even kill other animals, even people. Deadly animals use their bodies in different ways. Some have sharp **horns**. Some use their bite to hurt attackers and keep them away. Others use special parts of their body to kill for food.

Some animals kill other animals that come too close to them or their young. Many of the world's deadliest animals kill only in **defense**. Deadly animals are interesting animals. They can be found all over the world. Deadly animals will surprise you with their different shapes, sizes, and **abilities**!

The Cape buffalo is one of the deadliest animals in Africa. This animal uses its sharp horns to fend off attacks and to fight its fellow buffalo.

Killer Body

The deadliest animals use their body for **survival**. A strong, deadly body can help an animal get its food or fight off enemies. Some animals have sharp teeth for biting. Other animals have big horns that can cut and rip.

Sharp teeth and strong horns are not the only things that make an animal deadly. Many small animals have a different kind of strength. They use powerful **poison** against their enemies. These animals bite or **sting** other animals when they feel **threatened**. For example, the bite of the Sydney funnel-web spider can kill a person in 15 minutes.

The redback spider lives in Australia. It is a shy spider, whose poisonous bites can cause people pain, weakness, and sometimes even death.

Small but Deadly

To defend or feed itself, a small animal may bite or sting another. Deathstalker scorpions, for example, use poison to stop **prey** from moving. Then they eat the prey. When a bigger animal goes near a deathstalker, the scorpion may sting to defend itself.

Some deadly animals do not need to bite to kill. The poison dart frog's skin is so poisonous that most **predators** will not eat this frog. The deadliest poison dart frog is the golden poison dart frog. This animal lives in Central America. One golden poison dart frog has enough poison to kill 20,000 mice or 100 people!

There are many kinds of poison dart frogs and they come in many colors. This is a blue poison dart frog.

Killer Sharks

One of the deadliest animals in the ocean is the bull shark. Bull sharks are large and have a thick body. Male bull sharks can be 7 feet (2 m) long. They can weigh almost 200 pounds (90 kg). Female bull sharks can be even larger, at 11 ½ feet (3.5 m) and more than 500 pounds (227 kg)!

Bull sharks like to spend their time in water that is not too deep. When people or land animals go near the **territory** of bull sharks, the sharks may attack. Sharks' teeth are sharp and their mouth can be very large. A shark may grow more than 20,000 teeth in its lifetime!

Bull sharks live in warm water near coasts. These sharks are known as aggressive, or likely to attack.

Snakes and Snakebites

The black mamba is one of the world's deadliest animals. Though it can kill, a black mamba will try to run away from a predator if it can. Black mambas are some of the fastest snakes in the world. When deadly snakes do bite, though, they can kill quickly. The black mamba's poison can kill a person in 20 minutes!

The king cobra is another of the world's deadliest snakes. These large snakes can be fast and aggressive attackers. The poison from one bite of a king cobra can kill 20 people!

It is important to act quickly to treat a snakebite. One bite from a black mamba can have enough poison in it to kill up to 20 people!

One of the few animals that can kill a king cobra is the mongoose. The mongoose is a small furry mammal that lives in Africa and Asia.

The King Cobra

Powerful Bite

The king cobra is so poisonous that it can kill an elephant with a single bite!

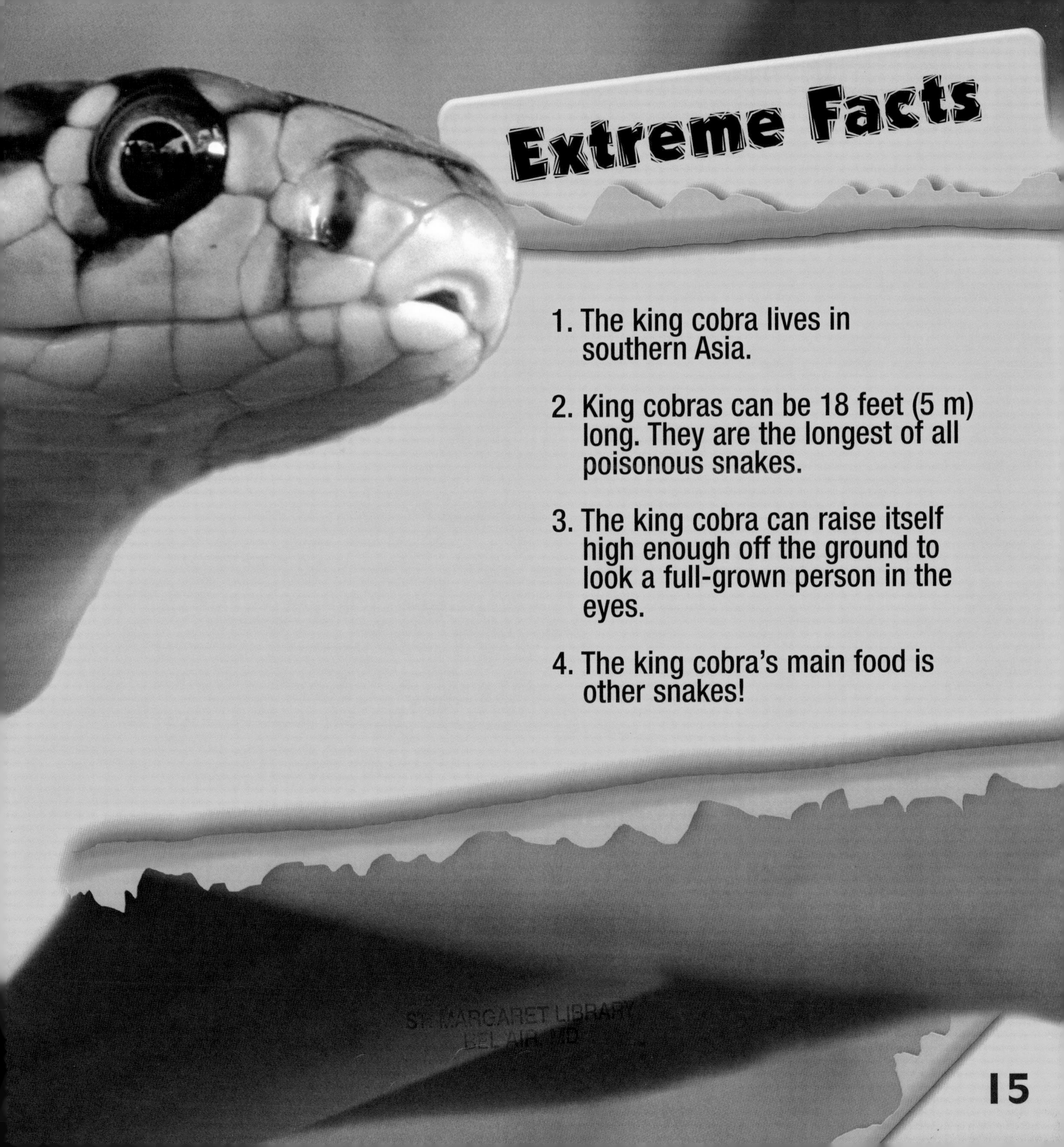

Extreme Facts

1. The king cobra lives in southern Asia.
2. King cobras can be 18 feet (5 m) long. They are the longest of all poisonous snakes.
3. The king cobra can raise itself high enough off the ground to look a full-grown person in the eyes.
4. The king cobra's main food is other snakes!

Large and Deadly

The world's largest reptile is the saltwater crocodile. Reptiles are animals such as turtles, snakes, and lizards. Reptiles are in the same animal class as dinosaurs. Saltwater crocodiles are sometimes called salties. They live in oceans, rivers, and wetlands. Salties are huge. They can be 20 feet (6 m) long and weigh 2,205 pounds (1,000 kg)!

With their huge, strong mouth and as many as 60 teeth, saltwater crocodiles can take giant bites. Some salties may be able to eat animals as large as buffalo. Saltwater crocodiles have been known to eat people, too!

The saltwater crocodile lives in Southeast Asia and Australia. It kills prey by biting and then rolling over it as it drags the prey into the water. This is called the death roll.

The Blue-Ringed Octopus

The blue-ringed octopus is the world's most poisonous octopus. One octopus can have enough poison to kill more than 20 people in minutes! This octopus is very small. Its body is only about 2 inches (5 cm) long.

The blue-ringed octopus has two kinds of poisons. It uses one kind of poison to kill prey, such as shrimp and crab. It uses another kind of poison to defend itself from predators.

When a blue-ringed octopus feels threatened, it will change color. Small blue rings show on its body only when the octopus is scared.

The blue-ringed octopus lives in warm ocean water near coasts. It lives in Australia, New Guinea, Indonesia, and the Philippines.

Natural Strengths

Deadly animals kill to survive. This can mean they kill to find food or to defend themselves. Without their giant mouth, which has many teeth, crocodiles could not eat. Small poison dart frogs have poison to guard them from being eaten by most predators.

Most people will never run into a cobra or a crocodile. However, learning about deadly animals from books or seeing them in the zoo can be just as exciting as seeing them in the wild. It is also much safer!

The lion is fast, strong, and has a powerful bite. All these things make it one of nature's deadliest animals.

Killer Facts

- **Mosquitoes** are the world's deadliest animal. Mosquito bites can carry an illness called malaria. Malaria kills millions of people all over the world every year.
- **Polar bears** can weigh 1,600 pounds (726 kg). Polar bears can kill elephant seals that weigh thousands of pounds (kg).
- **Sharks** have as many as 3,000 teeth at one time!
- **Box jellyfish** can have as many as 60 **tentacles**. Each tentacle has enough poison to kill 60 people.

Glossary

abilities (uh-BIH-luh-teez) Powers to do something.

defense (dih-FENTS) Something a living thing does that helps keep it safe.

horns (HORNZ) Hard parts on an animal's head.

poison (POY-zun) Something that causes pain or death.

predators (PREH-duh-terz) Animals that kill other animals for food.

prey (PRAY) An animal that is hunted by another animal for food.

sting (STING) To cause pain using a sharp part to hurt another animal.

survival (sur-VY-val) Staying alive.

tentacles (TEN-tih-kulz) Long, thin growths on animals that are used to touch, hold, or move.

territory (TER-uh-tor-ee) Land or space that animals guard for their use.

threatened (THREH-tund) Acted like something will possibly cause hurt.

Index

Web Sites

Due to the changing nature of Internet links, PowerKids Press has developed an online list of Web sites related to the subject of this book. This site is updated regularly. Please use this link to access the list:
www.powerkidslinks.com/exan/dead/